INDUSTRY STANDARD
OF THE PEOPLE'S REPUBLIC OF CHINA

Technical Standard for Structural Health Monitoring of Railway Passenger Station

TB/T 10184-2021

Prepared by: Shijiazhuang Tiedao University
Approved by: National Railway Administration of the People's Republic of China
Effective date: June 1, 2021

China Railway Publishing House Co., Ltd.

Beijing 2022

图书在版编目(CIP)数据

铁路客站结构健康监测技术标准:TB/T 10184-2021:
英文/中华人民共和国国家铁路局组织编译.—北京:
中国铁道出版社有限公司,2022.12
ISBN 978-7-113-29796-1

Ⅰ.①铁… Ⅱ.①中… Ⅲ.①铁路车站-客运站-建筑结构-安全监测-技术标准-中国-英文 Ⅳ.①TU248.1-65

中国版本图书馆 CIP 数据核字(2022)第 203910 号

Chinese version first published in the People's Republic of China in 2021
English version first published in the People's Republic of China in 2022
by China Railway Publishing House Co., Ltd.
No. 8, You'anmen West Street, Xicheng District
Beijing, 100054
www. tdpress. com

Printed in China by Beijing Forspeed Media Culture Co., Ltd.

© 2021 by National Railway Administration of the People's Republic of China

All rights reserved. No part of this publication may be reproduced or transmitted in any form or by any means, electronic or mechanical, including photocopying, recording, or by any information storage and retrieval systems, without the prior written consent of the publisher.

This book is sold subject to the condition that it shall not, by way of trade or otherwise, be lent, resold, hired out or otherwise circulated without the publisher's prior consent in any form of binding or cover other than that in which it is published and without a similar condition including this condition being imposed on the subsequent purchaser.

ISBN 978-7-113-29796-1

About the English Version

To promote the exchange and cooperation in railway technology between China and the rest of the world, National Railway Administration organized the translation of this Standard.

This Standard is the official English version of TB/T 10184-2021. The Chinese version of this Standard was issued by National Railway Administration and came into effect on June 1, 2021. In case of discrepancies between the two versions, the Chinese version shall prevail. National Railway Administration owns the copyright of this English version.

The Planning and Standard Research Institute of National Railway Administration and China Railway Construction Corporation (International) Limited prepared the English version. China Railway Economic and Planning Research Institute Co., Ltd provided great support during review of this English version.

Your comments are invited and should be addressed to the Technology and Legislation Department of National Railway Administration, No. 6 Fuxing Road, Beijing, 100860, P. R. China and the Planning and Standard Research Institute of National Railway Administration, No. 1, Guanglian Road, Xicheng District, Beijing, 100055, P. R. China.

Email: ghbzy_jsbzs@163.com

The translation was performed by Xiang Shang, Su Yujin and Zhang hongping.

The translation was reviewed by letter by Yu Bing, Zhang Hao, Yang Quanliang, Wang Lei, and Jing Guoqing.

Notice of National Railway Administration on Issuing the English Version of Three Railway Standards including *Code for Design of Cable-stayed Railway Bridge*

No. 2

The English version of three railway standards including *Code for Design of Cable-stayed Railway Bridge* (TB 10095-2020) is hereby issued. In case of discrepancies between the Chinese version and the English version, the former shall prevail.

China Railway Publishing House Co., Ltd. is authorized to publish the English version of these standards.

List of the English Version of Three Railway Standards

S/N	Title	Reference No.
1	*Code for Design of Cable-stayed Railway Bridge*	TB 10095-2020
2	*Unified Standard for Railway Information Modeling*	TB/T 10183-2021
3	*Technical Standard for Structural Health Monitoring of Railway Passenger Station*	TB/T 10184-2021

National Railway Administration

July 20, 2022

Notice of National Railway Administration on Issuing Railway Industry Standard
(Engineering and Construction Standard Batch No. 2, 2021)

Document GTKF [2021] No. 10

Three railway standards, including the *Unified Standard for Railway Information Modeling* (TB/T 10183-2021), the *Technical Standard for Structural Health Monitoring of Railway Passenger Station* (TB/T 10184-2021), and the *Technical Specification for Safety Monitoring of Operating Railway Infrastructures with Adjacent Constructions* (TB 10314-2021), will come into effect from June 1, 2021. China Railway Publishing House Co., Ltd. is authorized to publish these standards.

National Railway Administration
March 10, 2021

Foreword

With the rapid development of railway construction in China, many innovative large-scale complex structures are used for newly constructed modern railway passenger stations. Structural health monitoring and performance assessment of the railway stations under service conditions could bring the important benefits for guaranteeing the safety of operating such public venues with large crowds. In recent years, rich experience has been accumulated in the research and development and engineering application of structural health monitoring technology in large railway passenger stations, laying a solid foundation for the development of technical standard for structural health monitoring of railway passenger station.

This Standard is developed to unify the technical requirements of the structural health monitoring of railway passenger stations and raise the management efficiency in the design, installation and operation of the health monitoring system. The development is based on systematic analysis of the structural characteristics and safety risks of railway passenger stations, comprehensive summary of the research results and construction experience of the structural health monitoring technology of railway passenger stations in China, and relevant standards at home and abroad.

This Standard consists of 12 chapters, including General Provisions, Terms, Basic Requirements, Monitoring Parameters and Layout of Monitoring Points, Sensor Selection and Technical Requirements, Design of Data Acquisition and Transmission Module, Design of Data Management and Analysis Module, Design of Warning and Assessment Module, Installation and Protection of the Monitoring Equipment, System Integration and Debugging, System Acceptance, and System Maintenance.

The main contents in this Standard include the following:

1. Defines the basic principles and general requirements for structural health monitoring.

2. Specifies the monitoring object, parameters, and layout of the monitoring points.

3. Raises the requirements on the selection principle, technical index, and data acquisition mode of sensors.

4. Specifies the determination method on data acquisition configuration and data transmission mode.

5. Specifies the functions, technical requirements and implementation of the data storage, processing, management and analysis module.

6. Specifies the principles of structural safety warning threshold and structural condition assessment.

7. Specifies the installation and protection requirements of sensors, acquisition equipment, data cables, power cables, embedded parts, auxiliary facilities and monitoring center equipment.

8. Specifies the work contents and specific technical requirements for hardware and software integration and debugging of the monitoring system.

9. Puts forward the acceptance contents, methods and relevant requirements of monitoring system hardware, system software and functional modules.

10. Emphasizes the management requirements, maintenance methods and solutions of system maintenance.

We would be grateful if anyone finding the inaccuracy or ambiguity while using this Standard would inform us and address the comments to Shijiazhuang Tiedao University (No. 17, East Road of North 2nd Ring, Shijiazhuang, 050043) and copy to Planning and Standard Research Institute of National Railway Administration (No. 1, Guanglian Road, Xicheng District, Beijing, 100055) for reference in future revision.

The Technology and Legislation Department of National Railway Administration is responsible for the interpretation of this Standard.

Prepared mainly by:

Shijiazhuang Tiedao University

And also by:

China Railway Design Corporation Co., Ltd.

China Railway Siyuan Survey and Design Group Co., Ltd.

China Railway Construction Group Co., Ltd.

Drafted by:

Du Yanliang, Dang Li, Sun Baochen, Zhang Hao, Dong Cheng, Pan Guohua, Qian Zengzhi, Jiang Rui, Dong Suge, Kang Xiaogang, Liang Bing, Huang Zuguang, Liu Zhanwei, Song Huaijin, Wu Yingjun, Wang Wei, Fu Liyong, Deng Hai, Li Yanhui, Gao Xiujian, Bao Hua, Zhou Daxing, Shen Zhaowu, Li Haiyun, Zhao Weigang, Sun Tao, Shen Lei, Han Feng, Yuan Wei, Gao Yang, Wei Yukun.

Reviewed by:

Zhou Deliang, Liu Yan, Han Zhiwei, Tan Yueren, Xue Jigang, Liu Xun, Sang Cuijiang, Cheng Huilin, Zhang Guihai, Yin Jianchun, Yang Huidong, Liu Ming, Li Jingxue, Liao Yubiao, Yue Qing, Liu Chuanping, He Xianjuan, Hu Weiguang, Wang Ying, Zhang Wenxue, Lei Zhidong, Wang Yu, Lan Yanqiang, Liu Bojun, Zhang Maolin, Wu Zhishen, Li Hongnan.

Contents

1 General Provisions ··· 1
2 Terms ··· 2
3 Basic Requirements ·· 3
4 Monitoring Parameters and Layout of Monitoring Points ······································ 4
   4.1 General Requirements ··· 4
   4.2 Main Structure of Station Building and Platform Canopy Roof ················· 4
   4.3 Track Floor Structure of Elevated Station Building ····································· 5
5 Sensor Selection and Technical Requirements ·· 7
   5.1 General Requirements ··· 7
   5.2 Loads (Actions) and Environmental Monitoring Sensor ····························· 7
   5.3 Structural Response Monitoring Sensor ··· 8
6 Design of Data Acquisition and Transmission Module ·· 9
   6.1 General Requirements ··· 9
   6.2 Data Acquisition ··· 9
   6.3 Data Transmission ··· 10
7 Design of Data Management and Analysis Module ··· 11
   7.1 General Requirements ··· 11
   7.2 Data Storage ·· 11
   7.3 Data Management ··· 11
   7.4 Data Processing and Analysis ··· 13
8 Design of Warning and Assessment Module ·· 14
   8.1 General Requirements ··· 14
   8.2 Warning of Structural Safety ·· 14
   8.3 Structural Condition Assessment ·· 14
9 Installation and Protection of the Monitoring Equipment ···································· 16
   9.1 General Requirements ··· 16
   9.2 Installation and Protection of the Sensor ··· 16
   9.3 Installation and Protection of Data Acquisition Equipment ······················ 17
   9.4 Installation and Protection of Data Cable and Power Cable ····················· 17
   9.5 Installation and Protection of Embedded Parts and Auxiliary Facilities ··· 18
   9.6 Installation and Protection of Equipment in Monitoring Center ··············· 18
10 System Integration and Debugging ·· 19
   10.1 General Requirements ··· 19

  10.2 Software Development, Integration and Debugging ················· 19
  10.3 Hardware Integration and Debugging ································ 20
  10.4 Monitoring System Integration and Debugging ····················· 20
11 System Acceptance ·· 22
  11.1 General Requirements ··· 22
  11.2 Sensor Module Acceptance Inspection ·································· 22
  11.3 Acceptance Testing of Data Acquisition and Transmission Module ············ 22
  11.4 Acceptance Testing of Data Management and Analysis Module ················ 23
  11.5 Acceptance Checking of Alarm and Assessment Module ················ 23
  11.6 System Acceptance Documents ··· 24
12 System Maintenance ··· 25
  12.1 General Requirements ··· 25
  12.2 Hardware Maintenance ·· 25
  12.3 Software Maintenance ··· 26
  12.4 Emergency Maintenance ·· 26
Words Used for Different Degrees of Strictness ··· 27

1 General Provisions

1.0.1 This Standard is formulated for the structural health monitoring of railway passenger stations. It is designed to unify the technical requirements, improve the design, installation and operation management efficiency of the structural health monitoring system, and ensure the operational safety of railway passenger stations.

1.0.2 This Standard applies to structural health monitoring of large and very large railway passenger stations and passenger stations with special and complex structures.

1.0.3 On the basis of meeting the monitoring requirements during the operational period, the construction monitoring and operation monitoring may be combined to achieve continuous in the long-term monitoring.

1.0.4 The structural characteristics, installation process, operation conditions and other factors shall be comprehensively considered in the structural health monitoring scheme of railway passenger stations. The scheme shall be target-oriented, pertinent, and practical.

1.0.5 A structural health monitoring system of railway passenger stations shall be stable, reliable, technologically advanced, economically reasonable and convenient to maintain.

1.0.6 In addition to this Standard, the structural health monitoring of railway passenger stations shall also comply with relevant national standards in force.

2 Terms

2.0.1 Structural health monitoring

The process of collecting the on-site real-time structural response and environmental information by Non Destructive Testing (NDT), analyzing and assessing the health state of the structure, and giving an alarm to the abnormal state.

2.0.2 Operation monitoring

An activity to monitor the static and dynamic response of the structure under the environment and loads in the operational stage, obtain the structural condition information, assess its service condition and give an alarm to the operator, in order to ensure the normal use, safety and reliability of the structure.

2.0.3 Monitoring system

A system composed of sensors installed on the structure and the hardware and software used for data acquisition, transmission, analysis and processing, which are used to collect, process and analyze the loads, environmental impacts and structural response parameters of the structure, and assess the normal service level and safety state of the structure, and give alarms as needed.

2.0.4 Warning of safety

A warning of safety is to monitor the changes of structural condition, judge its abnormal states, and send different grades of warning information in time according to the degree of damage.

2.0.5 Warning threshold

A limiting value for alarming abnormal state of structure.

2.0.6 Condition evaluation/assessment

A set of assessments performed to assess the serviceability, safety, reliability and durability of certain members or the whole structure by using the condition assessment theory and method on the basis of structural monitoring data.

2.0.7 Monitoring system management and maintenance

The process of daily management, routine inspection, fault repair and other maintenance management to maintain the normal operation of the structural health monitoring system.

2.0.8 Viaduct-building integrated structure

The railway passenger station integrating the viaduct and the station building, which bears both train loads and building loads.

2.0.9 Long-span roof structure

Generally, it refers to steel structure roofs such as space grid truss, single-layer or double-layer reticulated shell, three-dimensional truss and cable structure with a span longer than 60 m; or solid web steel beam roof structure with a span longer than 30 m; or steel roof structure with a cantilever longer than 20 m.

2.0.10 Synchronous acquisition

All data acquisition channels of a data acquisition device start data acquisition at the same time.

3 Basic Requirements

3.0.1 A structural health monitoring system shall include sensor module, data acquisition and transmission module, data management and analysis module, alarm and assessment module. All the modules shall be coordinated and integrated.

3.0.2 The hardware of a monitoring system should be the products provided with mature technology and shall meet the requirements of stability, maintainability and durability.

3.0.3 The software of the monitoring system shall match the hardware and be compatible, extendable and easy to maintain.

3.0.4 The design scheme of a structural health monitoring system shall be determined according to the main structural characteristics of the passenger station, and shall include the following main contents:

    1 Monitoring parameters and layout of monitoring points.
    2 System structural and functional design of each module.
    3 Equipment selection, installation and protection.
    4 Data acquisition, transmission, management and analysis.
    5 Warning of safety and its assessment.

3.0.5 The installation of a structural health monitoring system shall not damage the anti-corrosion, fire prevention and other protective measures of the passenger station structures.

3.0.6 The hardware facilities in the monitoring system shall be effectively protected to ensure stability and reliability. The maintenance activities of the monitoring equipment shall not impact the consistency and continuity of data.

3.0.7 The feasibility of post-maintenance shall be considered in the layout and installation of hardware equipment.

3.0.8 The structural health monitoring system shall include the safety measures of preventing illegal penetration.

4 Monitoring Parameters and Layout of Monitoring Points

4.1 General Requirements

4.1.1 The monitoring parameters shall be determined according to the operation safety risks, structural stress characteristics, structural deformation characteristics, structural vulnerability analysis, service environment characteristics and monitoring objectives of railway passenger stations.

4.1.2 The layout of monitoring points shall meet the requirements of safety alarm and assessment, follow the principles of priority, representativeness, economy and feasibility.

4.1.3 The following structures or components should be included in the object of the structural health monitoring of railway passenger stations:

 1 The main structure of the station building roof.
 2 The main structure of platform canopy.
 3 The track floor structure of elevated station building directly bearing train loads.
 4 Other key structural members, components and bearings.

4.1.4 The overall structural response monitoring, partial structural response monitoring, loads (actions) and environmental monitoring shall be included in the monitoring system of railway passenger stations, and shall comply with the following provisions:

 1 The overall structural response monitoring shall include structural deformation, structural dynamic characteristics, dynamic response, etc.
 2 The partial structural response monitoring shall include partial strain (stress) of components, cracks, internal forces of structure, etc.
 3 The monitoring of loads (actions) and environment should include wind, rain, snow, temperature, humidity, earthquakes, etc.

4.2 Main Structure of Station Building and Platform Canopy Roof

4.2.1 Roof structures should be monitored in the following conditions:

 1 Steel structures with a span longer than 60 m, such as space grid truss, single-layer or double-layer reticulated shell, three-dimensional truss and cable structure.
 2 Beam structures with a span longer than 30 m.
 3 Steel structures with a cantilever longer than 20 m.
 4 Special structures that need to be monitored.

4.2.2 The structural health monitoring parameters for the main structure of the station building and platform canopy roof shall be selected according to Table 4.2.2 in accordance with the characteristics of the specific structure.

Table 4.2.2 Structural Health Monitoring Parameters for Roof Structures

Roof structure	Strain	Deformation		Environment				Bearing		Acceleration
		Vertical	Horizontal	Wind	Temperature	Humidity	Snow	Displacement	Reaction	
Space grid truss structure	★	★	▲	★	★	○	○	○	○	▲
Reticulated shell structure	★	★	▲	★	★	○	○	▲	○	▲

Table 4.2.2(continued)

Roof structure	Strain	Deformation		Environment				Bearing		Acceleration
		Vertical	Horizontal	Wind	Temperature	Humidity	Snow	Displacement	Reaction	
Truss structure	★	★	▲	★	★	○	○	○	○	▲
Cable structure	★	★	▲	★	★	○	○	▲	○	★
Cantilever structure	★	★	▲	★	★	○	○	▲	○	★
Special structure	★	★	▲	★	★	○	○	▲	○	▲

Notes: 1 ★ items shall be monitored, ▲ items should be monitored, ○ items may be monitored.
   2 Cable tension monitoring shall be included in the monitoring of cable structure.
   3 Seismic monitoring may be included in areas with seismic fortification intensity of 8 and above.
   4 The displacement of bearings shall be monitored for non-fixed bearings.

4.2.3 The key components of the main structure of the roof or canopy shall be subject to strain monitoring. They mainly include the root of the cantilever component, the midspan of the simply supported component, the midspan and the bearing of the continuous component.

4.2.4 The positions of points, lines and surfaces, which can reflect the overall stress and deformation characteristics of the structure, shall be comprehensively considered in the layout of deformation monitoring points of the main roof structure.

4.2.5 The layout of vibration monitoring points for the station building roof structure shall be determined according to the structural dynamic characteristics, stress characteristics and other factors.

4.2.6 The vibration monitoring points for the platform canopy roof structure shall be arranged at the midspan of horizontal components above the passing area of trains on the main track.

4.2.7 For cable tension monitoring, typical components with large tension force or wide variation of tension force shall be selected according to the structural characteristics, and appropriate monitoring methods shall be selected according to the characteristics of tension components.

4.2.8 The anemometer should be set at a position with strong wind and no vortex on the roof of the station building.

4.2.9 The temperature monitoring points for structural members shall be determined according to the analysis of temperature effect and the temperature compensation requirements of strain monitoring points.

4.3 Track Floor Structure of Elevated Station Building

4.3.1 The track floor structure of the elevated station building should be monitored in the following conditions:
  1 Railway track beam with a span not less than 24 m.
  2 New type of track floor structure.

4.3.2 The structural health monitoring parameters for the track floor structure of the elevated station building shall be determined according to Table 4.3.2 in combination with the structural characteristics of the implemented project.

Table 4.3.2 Structural Health Monitoring for Track Floor Structure

Track floor structure	Strain	Reinforcement stress	Vertical deformation	Cracks	Vertical and horizontal acceleration
Railway track beam	★	▲	★	▲	▲

Notes: 1 ★ items shall be monitored, ▲ items should be monitored.
   2 Reinforcement stress monitoring shall include monitoring of ordinary reinforcement and prestressed reinforcement of special components. When conditions permit, the stress of prestressed reinforcement may be monitored.

4.3.3 The monitoring points of railway track beam should be arranged in areas with large internal force and complex stress, such as midspan, bearing.

4.3.4 The layout of vertical deformation monitoring points shall reflect the deformation characteristics of the structure, and a deflection curve shall be formed by the results of each monitoring point.

4.3.5 The layout of vibration monitoring points for the structure of railway track beam shall be determined according to structural dynamic characteristics, stress characteristics and other factors.

5 Sensor Selection and Technical Requirements

5.1 General Requirements

5.1.1 Sensors shall meet the requirements of sensing range, accuracy, resolution, sensitivity, frequency response characteristics, stability, durability, environmental adaptability, etc.

5.1.2 Sensors shall be selected based on the convenience of the site installation and system integration.

5.2 Loads (Actions) and Environmental Monitoring Sensor

5.2.1 For wind speed and direction monitoring, an anemometer shall be selected according to the regional or environmental conditions, and meet the following requirements:

 1 The range of the anemometer shall be greater than the maximum wind speed once in 100 years.

 2 An 8-direction anemometer should be selected. The start-up wind speed shall not be greater than 0.6 m/s.

 3 The accuracy of the anemometer should not be less than 0.1 m/s in speed and should not be less than 3° in direction.

 4 The operating ambient temperature of the anemometer is $-50\ ℃ - +70\ ℃$, and the relative air humidity is no higher than 90%.

5.2.2 For wind pressure monitoring, micro pressure sensors which can detect both positive and negative pressure shall be selected. In addition, specialized anemometers which use air pressure as monitoring parameter may be selected and shall meet the following requirements:

 1 The pressure range of the anemometers shall meet the requirements of the wind field in the structural design.

 2 The accuracy of the anemometers shall be ±0.4% of the full scale, and should not be less than 10 Pa. The nonlinearity shall be within ±0.1% of the full scale, and the response time shall be less than 200 ms.

5.2.3 The temperature monitoring sensors shall meet the following requirements:

 1 The temperature range of the sensors for monitoring the atmospheric ambient temperature should be 20 ℃ higher than the local maximum temperature and 20 ℃ lower than the local minimum temperature over the years.

 2 The temperature range of the sensors for monitoring the surface temperature of the structure should be 50 ℃ higher than the local maximum temperature and 20 ℃ lower than the local minimum temperature over the years.

 3 The accuracy of the atmospheric ambient temperature sensors should not be lower than ±0.5 ℃, and the resolution should not be lower than 0.1 ℃.

 4 The accuracy of the structural surface and the internal temperature sensors should not be lower than ±0.2 ℃, and the resolution should not be lower than 0.1 ℃.

5.2.4 Humidity sensors such as lithium chloride hygrometers, capacitive hygrometers and electrolytic hygrometers may be selected for humidity monitoring. The sensors shall be able to measure the full range of relative humidity, from 0 to 100%. The relative humidity accuracy should not be less than

3%. Other types of sensors may also be used after verification.

5.2.5 Strong motion seismographs or other verified sensors should be selected for seismic monitoring, and the following requirements shall be met:

1 The seismic monitors shall be able to eliminate the influence of atmospheric pressure and temperature changes on the instrument.

2 The seismic monitors shall be able to work stably in complex environment.

3 The seismic monitors shall be equipped with environmental protection devices with sealing, and shall be provided with high-precision constant temperature function.

5.3 Structural Response Monitoring Sensor

5.3.1 For overall or partial vibration monitoring of the structure, sensors shall be selected according to the requirements of overall dynamic calculation and analysis, environmental adaptability and durability. The following requirements shall be met:

1 The appropriate sensitivity, frequency response range and monitoring sensing range shall be obtained by the vibration sensor.

2 Three-dimensional, bidirectional or unidirectional vibration sensors may be selected according to the main vibration modes of the structure and the location of monitoring points.

3 For the long-span station building structure with low basic frequency, the vibration sensors with excellent low frequency performance should be selected. The range is not less than $\pm 2g$, the transverse sensitivity is generally 1% - 5% of the axial, the frequency response is 0.2 Hz - 1 500 Hz ($\pm 10\%$), and the service temperature range is $-50\ ^\circ\text{C} - +70\ ^\circ\text{C}$.

5.3.2 For displacement monitoring, the specific displacement sensor shall be selected according to monitoring objects. The monitoring sensing range shall be 10 mm - 1 000 mm, the sensitivity should not be less than 0.1 mm, and the accuracy should not be less than 0.5%F.S.

5.3.3 Strain sensors shall meet the following requirements:

1 High resolution, wide sensing range, well anti-fatigue, long service life and other basic functions should be provided. The range of static strain sensors should not be less than 1.5 - 2 times of the predicted maximum, and the range of dynamic strain sensors should not be less than 2 times of the predicted maximum.

2 Sensors are generally classified into static strain sensors and dynamic strain sensors, and temperature compensation shall be carried out for strain monitoring.

3 Sensors shall be able to work in various environments with high resistance to electromagnetic interference, chemical-corrosion, and shock and vibration.

4 The sensing range should not be generally less than $\pm 1\ 500\ \mu\varepsilon$; the accuracy should not be less than $\pm 0.1\%$F.S; the nonlinearity should be less than 0.5%F.S, and the sensitivity should be less than 1.0 $\mu\varepsilon$. In addition, the service temperature range is $-50\ ^\circ\text{C} - +70\ ^\circ\text{C}$.

5.3.4 The specification of the cable force sensors shall be selected according to the structural design, calculation and analysis, and the monitoring accuracy should not be less than 3%F.S.

5.3.5 The range of the crack monitoring sensor is generally between 1 mm - 20 mm, and the minimum division value of crack width monitoring should not be greater than 0.05 mm. For cracks with width less than 1 mm, the resolution of the sensor should not be less than 0.01 mm.

5.3.6 Bearings which can measure reaction forces should be selected for bearing reaction forces monitoring, and other types of sensors can also be used after verification.

6 Design of Data Acquisition and Transmission Module

6.1 General Requirements

6.1.1 The data quality and integrity shall be guaranteed during data acquisition and transmission.

6.1.2 The design of data acquisition and transmission module shall include designs such as the portals to match with the sensor interface, signal conditioning and data acquisition scheme, routing and data transmission scheme, software functions, integrations.

6.1.3 The data acquisition and transmission method shall match with the selected sensors in both hardware and software.

6.1.4 The data acquisition and transmission software shall reach automatic data acquisition and transmission, and the collected and transmitted parameters can be set and adjusted manually.

6.1.5 The data acquisition scheme shall include the determination for data acquisition mode, threshold trigger and sampling frequency.

6.2 Data Acquisition

6.2.1 There shall be a clear topological relationship between data acquisition equipment and sensors. Data acquisition method shall be decided according to the space size of railway passenger stations, number and layout of monitoring points and sensor type. Centralized, distributed or hybrid data acquisition method may be selected.

6.2.2 The data acquisition equipment shall be selected according to the requirements of sensor output signal type, adaptability, compatibility, sampling frequency, accuracy and resolution. When correlation analysis (including modal analysis) is required for same or different types of data, the relevant data shall be collected synchronously. The signal conditioning and signal/noise separation shall be performed for the signals collected by the data acquisition equipment to improve the quality of data acquisition.

6.2.3 For the data acquisition scheme, the methods of real-time sampling, time sampling, trigger sampling or mixed sampling can be selected according to the loads, environment, structural response characteristics and monitoring parameters.

6.2.4 The sampling frequency shall be set according to the monitoring parameters and functional requirements, which shall reflect the real state of the monitored structure and meet the application requirements of structural health monitoring data.

6.2.5 The self-calibration function should be used for data acquisition. If there is no self-calibration function, it shall be calibrated regularly.

6.2.6 Anti-interference measures, such as serial mode interference, common-mode interference, earthing technology and shielding technology may be adopted for data acquisition to improve the quality of signal acquisition.

6.2.7 The data acquisition station (DAS) shall be set according to the monitoring requirements and distance requirements of signal transmission without affecting the data quality. Time synchronization requirements of data acquisition shall be considered for data acquisition stations.

6.3　Data Transmission

6.3.1　The time delay of the monitoring system shall also be comprehensively considered while determining the data transmission scheme.

6.3.2　The hardware, data transmission and monitoring system modules shall be a dynamic integration to ensure the transmission of monitoring data and signals among modules to be real-time, reliable, uninterrupted, and to support breakpoint continuous transmission.

6.3.3　A wired transmission or verified wireless transmission should be adopted for data transmission. Necessary protective measures shall be taken for the transmission lines to avoid electromagnetic interference.

6.3.4　Industrial Ethernet, optical fibers, or other digital signal transmission methods should be used for wired transmission.

6.3.5　The short distance and long-distance wireless transmission methods such as RFID, ZigBee, WLAN, 4G and 5G should be selected for wireless transmission. Signal transmitters and receivers shall be kept away from strong electromagnetic interference sources.

6.3.6　Optical fiber transmission, wireless transmission or a combination of both should be used for the long-distance data transmission between the on-site collectors of the railway passenger stations and the monitoring center.

7 Design of Data Management and Analysis Module

7.1 General Requirements

7.1.1 The data management and analysis module shall include the following key functions:
1 Data preprocessing, data storage, automatic generation of reports.
2 Establishing a central database for data query and management.
3 Data backup, automatic and manual data import and export.
4 Unified data standard formats and interface, which are incorporated into the system software manual.
5 Technical measures such as setting different user-level permissions and passwords, as well as network firewall protection, shall be taken to improve data security.

7.1.2 The database of the data management and analysis module should be designed according to the type of data. The maintenance and backup mechanism should be designed for the database. Mature commercial database software systems or open-source database software systems that are widely applied should be selected for the data management and storage.

7.1.3 The design of the data management and analysis module shall include functional design, structural design, performance design, and security design. Factors such as security, stability, fault tolerance, operability and extendibility shall be taken into consideration in the software development.

7.1.4 Data storage and management should be performed on the local computer. Cloud storage and cloud management technologies may also be used.

7.1.5 Data preprocessing shall be carried out before data analysis to ensure the objectivity and integrity of the data.

7.2 Data Storage

7.2.1 The database design for data storage shall follow the principles of reliability, advancement, openness, extendibility, standardization and economy, and the shareability of data, the integrity of the data structure, the consistency between the database system and the application system shall be guaranteed.

7.2.2 The original monitoring data shall be regularly stored, backed up and archived. The hard disk capacity of the monitoring center shall support the original dynamic data store for not less than 3 months, and static data preservation for the full life cycle. The data obtained after processing and analysis shall be stored separately.

7.3 Data Management

7.3.1 The modular structure of database shall support to store and manage the structure information of passenger station, information of monitoring system and monitoring data in grades and categories, and should include sub-databases of passenger station structure information, monitoring system information, real-time data, statistical analysis data, structural safety assessment, etc.

7.3.2 The processed data should be stored in sufficient online storage for online query and analysis at any time. The statistically analyzed data should be stored separately. After quarterly or annual data analysis, a certain section or several sections of typical data should be stored. All of the data and visual images should be able to be reproduced and be displayed.

7.3.3 The sub-database for monitoring system information shall be able to store and manage the information such as sensors, data acquisition and transmission equipment, data processing and management equipment and software. Such information refers to equipment installation location, technical parameters, brand, specifications, and the full name and version number of the software involved.

7.3.4 The warning thresholds, the safety assessment methods, safety assessment results and alarm history records shall be stored and managed in the system database. They should be seamlessly connected and data shared with the passenger station routine inspection and the maintenance management system of the passenger station.

7.3.5 The data report function shall support generating quarterly reports, annual reports or special reports after particular events automatically. The generated reports shall be exported to a common document format that is easy to invoke by the office system.

7.3.6 Database management shall include the management such as monitoring equipment, monitoring information, structural model information, assessment and analysis information, data dump, users, security and alarm information.

7.3.7 Monitoring equipment management shall include the functions of adding, replacing, status querying and malfunctioning detection of sensors, acquisition equipment (including central stations and sub-acquisition stations), and signal conditioning equipment. Sensors should be classified and managed according to the contents and functions of monitoring information.

7.3.8 Monitoring information management shall include importing monitoring information automatically, exporting data to graphics or files, and querying historical monitoring information. The monitoring information shall be visualized.

7.3.9 The information management of passenger stations structural model shall allow the storage and query of basic structural parameters.

7.3.10 The management of information assessment and analysis shall support the storage and query of assessment criteria, assessment model, assessment results, etc.

7.3.11 The data storage shall support the archiving and management of original data. The archived data may be stored in the mass storage device and shall be accessible during use.

7.3.12 The user management shall support the definition and allocation of user permissions, different modules shall be operated according to the user permissions, and the role-based user group management, user authorization, registered account and authentication management shall be provided by the user management.

7.3.13 System security management shall contain the functions of network security management and protection of system operation environment, backup and disaster recovery of the database, sensitive information marking, usage log audit, etc. The security management of the database system shall be supported by corresponding hardware, software and personnel.

7.3.14 Data loading shall include the main steps such as data screening, input, verification, conversion and synthesis.

7.3.15 The response time grade of the query shall be seconds, and the data analysis and visualization shall be able to meet the use requirements.

7.4 Data Processing and Analysis

7.4.1 The monitoring data shall be preprocessed by means of cleaning, denoising, etc.

7.4.2 The data of loads and environment, and the data of overall and partial response of the structure shall be comprehensively analyzed to provide basic data for safety alarm, safety assessment and special assessment of the structure.

7.4.3 The scope of analysis should include basic statistical data analysis, nonlinear data-fitting analysis, predictive analytics and correlation analysis.

7.4.4 The data analysis for external influencing factors of passenger stations includes but not limited to wind load, temperature load, snow load, seismic action and air humidity.

    1 The wind load analysis should include wind speed, wind direction, wind attack angle, pulsating wind speed spectrum, turbulence intensity, gust factor, fatigue wind speed spectrums, etc.

    2 The temperature load analysis should include the maximum temperature, the minimum temperature, the maximum temperature gradient of the section, etc.

    3 The snow load analysis should include the maximum snow thickness, the minimum snow thickness, snowmelt cycle, etc.

    4 The seismic action analysis should include peak acceleration, peak velocity, earthquake duration, seismic signal spectrum, earthquake response spectra, etc.

    5 The air humidity analysis should include absolute air humidity, relative air humidity, critical relative air humidity, etc.

7.4.5 The data analysis of structural response of passenger stations shall include but not limited to deformation, displacement, stress, etc.

    1 The data analysis of monitored deformation and displacement shall include the analysis of average value, absolute maximum, and deformation and displacement direction. The cumulative value and rate of change should be analyzed.

    2 The maximum and root-mean-square of acceleration shall be analyzed, and the correlation analysis of structural vibration and external load should be carried out.

7.4.6 The data analysis report shall include but not limited to the following contents:

    1 Project overview.

    2 Method and principle of data analysis.

    3 Layout of monitoring points and equipment.

    4 Monitoring period and duration.

    5 Data status and comparative analysis.

    6 Alarm and prediction based on monitoring data.

    7 Environmental impact and risk analysis.

    8 Structural condition assessment.

    9 Suggested measures and actions.

8 Design of Warning and Assessment Module

8.1 General Requirements

8.1.1 For the warning and assessment module, structural hierarchical safety warning and structural condition assessment shall be carried out according to the structural monitoring objectives and requirements.

8.1.2 For warning of structural safety, the influence of service conditions on structural health shall be comprehensively analyzed, and hierarchical index of warning of safety shall be set.

8.1.3 For structural condition assessment, a comprehensive analysis shall be carried out based on the monitoring data to accurately reflect the current performance of the structure, and corresponding structural maintenance and management suggestions shall be put forward in time.

8.2 Warning of Structural Safety

8.2.1 The warning of structural safety is graded into three levels: yellow, orange and red. The warning may also be graded into more levels according to specific conditions.

8.2.2 The warning threshold for structural safety shall be determined according to the comprehensive analysis on design requirements, initial status information of the structure, monitoring data and data analysis.

8.2.3 The yellow warning threshold should be set according to the design value of internal forces or deformation of structural members, nodes, and bearings. The red warning threshold should be determined according to the structural status when slight damage happens. The orange warning threshold may be determined by the middle value of yellow and red warning thresholds.

8.2.4 The warning of structural safety should be automatic and real-time. Various warning information should be sent to relevant personnel in time as needed.

8.3 Structural Condition Assessment

8.3.1 If one of the following conditions occurs in the structure, the structural condition shall be assessed.

  1 It is found that the structure is abnormal.

  2 Alarm occurs in the process of structural inspection, detection or monitoring, unclarified obvious structural defects are found, or the existing structural defects show an accelerated development trend.

  3 The structure has experienced extreme weather.

  4 The structure has encountered an emergency.

  5 The scope of function or usage of the structure has been changed.

8.3.2 The structural condition assessment based on structural health monitoring data shall be conducted according to certain criteria. According to chosen criteria, the specific assessing method can be classified as threshold assessment, reasonable value or reasonable scope assessment, data assessment at a reference time point (such as on completion), or condition comparison assessment in different stages. In the structural condition assessment, based on monitoring data, one or several methods may be used to assess the condition of the structure.

8.3.3 The fatigue assessment for steel structure by using strain sensor shall comply with the following provisions:

1 Fatigue assessment may be omitted for components only under pressure.

2 Allowable Stress or Fatigue Damage Index should be used to assess the fatigue state of components at the monitoring points.

3 Allowable Stress should be used to assess the fatigue state of the structure.

4 Rain Flow Counting and Miner Rule for cumulative fatigue damage shall be adopted to calculate the cumulative fatigue damage index D of structural members at the monitoring point, and the fatigue assessment shall be carried out according to the provisions of Table 8.3.3.

Table 8.3.3 Grades of Fatigue State

D	Status of monitoring point on component
0 - 0.05	Sound condition
0.05 - 0.20	Good condition
0.20 - 0.45	Moderate damage condition
0.45 - 0.80	Severe damage condition
>0.80	Hazardous condition

Note: The effect of corrosion on fatigue life is not considered in the fatigue state grades shown in this table. When corrosion occurs, the adverse effect of corrosion on fatigue life of steel members shall be considered.

9 Installation and Protection of the Monitoring Equipment

9.1 General Requirements

9.1.1 The monitoring equipment shall be installed safely and reliably. The warning signs shall be stable and conspicuous.

9.1.2 The monitoring equipment shall be protected against collision, lightning, water and humidity, interference, high and low-temperature to avoid accidental damage and ensure its stable and reliable operation.

9.1.3 The installation of monitoring equipment shall be easy to maintain.

9.1.4 The power supply of monitoring equipment shall be stable and reliable, and an UPS shall be equipped.

9.1.5 The installation and protection plan should be integrated in the detailed design of related services.

9.1.6 The sensor shall be calibrated or self-calibrated before installation.

9.2 Installation and Protection of the Sensor

9.2.1 The sensors and monitoring points shall be numbered. The installation direction of sensors shall be clear. When a sensor is fixed on the surface of a structure, they should be welded or bolted to the surface, and protective measures shall be taken.

9.2.2 The external wire connection of the sensor shall meet the following requirements:

　1　The wire arrangement shall be neat and orderly, and the conductors shall be well insulated and free of damage.

　2　Anti-corrosion bolts shall be used for fixing wire.

　3　The bolts for fixing the wires shall be tightened, and the tightening torque shall meet the requirements of the specification.

　4　The tension of the wires shall be moderate to avoid overstress on the inside of the sensors.

9.2.3 During installation, the deformation sensors and displacement sensors shall be leveled, and the ambient temperature and humidity shall meet the design requirements.

9.2.4 When installing acceleration sensors, the vibration sensitive axis of the sensor shall be guaranteed to be aligned with the monitored vibration direction, and the surface of the sensor shall be smooth. The sensors and the monitored components shall be rigidly connected. The impedance between the earthing shell and the earthing shall be less than 1 Ω.

9.2.5 The effectiveness of strain sensors shall be confirmed one by one before installation to ensure normal operation. The deviation of the installation position in all directions shall not be greater than 30 mm from the monitoring point, and the deviation of installation angle shall not be greater than 2°. The sensors should be welded or bolted to the monitored objects.

9.2.6 The layout of monitoring sensors for wind parameter shall be free of obstructions in all directions. The types of sensors shall be the same. The structure which sensors installed on shall have sufficient strength and stiffness.

9.2.7 The strong motion seismograph shall be installed on a dedicated base with pre-embedded anchor bolts to be placed at the center. The base surface shall be smooth and flat.

9.2.8 The crack monitoring sensor shall be installed perpendicular to the monitored crack, and the vertical deviation shall not be greater than 1°. The axis of the crack gauge shall be parallel to the surface where the crack is located, and the parallelism deviation shall not be greater than 1°.

9.2.9 The installation of the sensors for bearing reaction forces monitoring shall not change the building elevation and the connection mode between the structure and the bearings.

9.3 Installation and Protection of Data Acquisition Equipment

9.3.1 The signal acquisition equipment shall be installed in the environment with little interference. If it is unavoidable, effective shielding measures shall be taken.

9.3.2 The acquisition equipment should be installed at the position where the transmission line is short and the signal loss is minimum.

9.3.3 The installation of acquisition equipment shall be stable, reliable, and meet the requirements of relevant specifications for equipment installation.

9.3.4 The acquisition equipment shall be replaceable so as to facilitate system maintenance.

9.3.5 Protective measures for acquisition equipment shall be taken as indicated in the manual. The metal door of the electrical control cabinet shall be locked. The protective barriers and warning signs shall be set if conditions permit.

9.4 Installation and Protection of Data Cable and Power Cable

9.4.1 Data cables and power cables may be placed in cable containments or fixed with cable clips. The common cable containments of the building shall be prioritized used, and data cables and power cables shall be marked distinctly. The antenna of wireless equipment should be installed inside the FRP cabinet. If metal cabinet is used, the antenna shall be placed on the top of the metal cabinet and shall not be obstructed.

9.4.2 Data cables and power cables shall be protected against lightning, static electricity, dust, water, etc. The protective device shall be reliable and durable, and the color shall be in harmony with the environment and not affect the overall aesthetics of the building.

9.4.3 When laying data cables and power cables, protective measures shall be taken according to the requirements of design documents. When there is no such requirements in the design document, soft corrugated pipe or steel pipe may be used for protection. The data cables and power cables shall not be laid at positions which may affect the operation and maintenance of equipment and pipelines, or the locations which may affect the transportation of vehicles and pedestrians. The transportation channel, pedestrian passage and lifting eyes of equipment shall be avoided. For cabinets and racks exposed to rain and direct sunlight, cables should pass through the bottom of cabinets or racks with waterproof measures.

9.4.4 Data cables and power cables shall not be laid in areas with corrosive substance, strong magnetic and electric field interference, and protective or shielding measures shall be taken if it is unavoidable.

9.4.5 Surplus length of the cable shall be kept at cable terminals, expansion joints and settlement joints.

9.4.6 No intermediate joints shall be made in the cable. If unavoidable, joints shall be placed in the junction box. Crimping shall be used for cable joint. When cables are welded, non-corrosive flux shall be used. Crimping shall be used for compensating cable.

9.4.7 Intermediate joints shall be avoided in the optical fiber cables. If unavoidable, special tools shall be used to make fusion splicing, and the number of splicing joints shall be minimized. The splice loss shall be less than 0.1 dB.

9.5 Installation and Protection of Embedded Parts and Auxiliary Facilities

9.5.1 Before installation of embedded parts and auxiliary facilities, the installation and protection scheme shall be developed according to the monitoring scheme, installation location and environment.

9.5.2 The embedded parts for fixing the sensor shall not be omitted, and shall be installed firmly in the correct location. The deviation allowance between the center-line of embedded parts and the installation axis shall be limited to ± 3 mm.

9.5.3 The strain gauge and exposed welding joint shall be protected by moisture-proof insulation adhesive. The data cable shall be functioned with shielding and be insulated from the monitored object. The data cables of different types of sensors shall be classified, compactly arranged and protected.

9.5.4 The installation and acceptance of lightning protection devices, earthing devices and earthing wires shall comply with the requirements of the *Technical Code for Protection of Building Electronic Information System Against Lighting* (GB 50343) and meet the following provisions:

  1 The lightning protector shall be firmly installed and reliably wired. The installation position and sequence of multiple lightning protectors shall meet the requirements of design documents and product manuals.

  2 The earthing wires shall be set separately without using the existing lightning protection wire of the station building structure. The embedded position and depth of the earth electrode shall meet the requirements of design documents.

  3 Physical or chemical treatment shall be done to reduce the resistance when the earthing resistance cannot meet the requirements of design documents.

9.6 Installation and Protection of Equipment in Monitoring Center

9.6.1 The equipment in monitoring center shall be installed in a quiet, dry and clean environment without static electricity, strong magnetic field and high temperature.

9.6.2 The equipment in monitoring center shall be far away from areas producing dust, oily fume, harmful gas, flammable, explosive and corrosive substances.

9.6.3 The equipment in monitoring center shall be neat, and the cables shall be embedded and all electrical equipment shall be placed in protective cabinets. There shall be no water supply or drainage pipes above the equipment.

9.6.4 The monitoring center shall be provided with protection functions such as heat dissipation, fire prevention and dust prevention, and the equipment installation position shall be ventilated.

9.6.5 The internal temperature, humidity and other conditions of the monitoring center shall meet the work environment requirements of communication equipment and other equipment.

9.6.6 The hardware equipment in monitoring center shall be firmly fixed to prevent system breakdown caused by hardware movement or overturning due to earthquake.

9.6.7 The installation and acceptance of monitoring center may be carried out in accordance with the relevant requirements of the *Code for Acceptance of Quality of Intelligent Building Systems* (GB 50339), and meet the relevant requirements of the *Code for Design of Data Centers* (GB 50174).

10 System Integration and Debugging

10.1 General Requirements

10.1.1 The system integration and debugging shall include the integration and debugging of hardware, software and platform tools.

10.1.2 The system integration and debugging shall optimize the configuration of hardware and software, which in turn optimize the overall performance of the system.

10.1.3 The testing interface for integration and debugging shall be reserved for both software and hardware during system design and development.

10.1.4 The debugging plan shall be developed, and necessary safety protection measures shall be taken before system debugging.

10.1.5 The system shall be debugged according to the design documents and the product technical manuals.

10.1.6 The following preparations shall be made before system debugging:

1 The specification, model, quantity and marks of installed hardware and software shall be checked against the installation records to ensure the correctness.

2 The earthing system and earthing resistance shall be checked whether they meet the design requirements. While there is no specific requirements in design, the earthing resistance shall be less than 4 Ω.

3 The insulation resistance value between different earthing systems shall be greater than 50 MΩ when the common connection points are disconnected.

4 The input power supply shall be checked with the design or equipment power supply requirements, including power supply type, voltage, load capacity, etc.

10.1.7 The system debugging includes single system debugging and integrated system debugging, and the debugging report shall be developed.

10.1.8 The measuring and testing instruments used for system verification shall be calibrated and within the specified validity period, and the accuracy of instruments shall not be lower than that of sensors to be verified.

10.2 Software Development, Integration and Debugging

10.2.1 The system software should be developed according to the unified software platform, with extendibility, easy maintenance and good service performance.

10.2.2 The user interface of system software shall be well designed and easy to use.

10.2.3 The user interface of system software shall include the entrance or display interface for static and dynamic information of passenger stations such as basic information, system information, monitoring data, data acquisition and transmission, data processing and management, safety alarm and assessment.

10.2.4 The system software shall be compatible with multiple types of databases.

10.2.5 The system software shall be able to display the alarm information in real time and transmit the information to the designated system or personnel of the passenger stations.

10.2.6 The system software shall be able to display in real time the equipment status information such as sensing, acquisition, transmission and data processing, and shall be able to allow online setting and modification of the functional parameters of each module.

10.2.7 The system software shall provide access to the data analysis module, and provide interfaces for storing, calling or displaying various assessment results and reports in the user interface.

10.2.8 The network security measures such as hierarchical authorization shall be used to allow users to log in remotely through internet encrypted tunnel and to view monitoring data and reports online.

10.2.9 After the completion of the system software, software stability and functions such as software interface, data acquisition, transmission, storage, query, analysis and processing, alarm and assessment, shall be tested. The bugs found during testing and the solutions shall be recorded. The debugging report shall be prepared after the debugging meets the requirements.

10.3 Hardware Integration and Debugging

10.3.1 The monitoring equipment includes various sensors, the master data acquisition devices and the slave data acquisition devices. When designing the performance and function of the hardware, a targeted type selection shall be carried out according to the overall requirements of the monitoring system and the actual scenarios on-site.

10.3.2 The acquisition equipment shall be compatible with the sensor and meet the requirements of the monitored index. When the monitored index is the structural dynamic response signal, the sampling frequency of the acquisition equipment shall meet the frequency requirements of the required monitored index.

10.3.3 When correlation analysis, including modal analysis is required for similar or different types of data, all relevant data shall be acquired synchronously.

10.3.4 After the installation of hardware, the equipment operation status, stability of signal transmission, synchronization and reliability of collected data, etc. shall be tested. Bugs found during testing shall be removed in time and recorded. After the debugging meets the requirements, the debugging report shall be prepared.

10.4 Monitoring System Integration and Debugging

10.4.1 The system integration shall meet the overall requirements of the monitoring system, and each subsystem and module shall be compatible, reliable for data transmission, stable, adaptable for environment and extendable.

10.4.2 The system integration shall meet the requirements of network communication, environmental adaptability and lightning protection, as well as the compatibility and adaptability with data acquisition interface, communication interface, data protocol and power supply interface. It shall reach reliability and stability with optimal allocation and maximized reliability as the constraint condition.

10.4.3 The temperature, humidity, power supply and electromagnetic interference in the monitoring system equipment room shall meet the normal operation conditions of the equipment. UPS, air conditioners, lightning protection and electromagnetic interference prevention equipment shall be provided.

10.4.4 The user interface shall clearly display the monitoring point position, monitoring data, work status of data acquisition and transmission, data processing and analysis results, safety alarm information and assessment. The functional parameters of each module shall be able to be set and modified online.

10.4.5 In system integration, the security of the hardware and software system, especially the information security of the software system shall be guaranteed.

10.4.6 The overall structure of system integration shall be distributed, open, extendable and nonexclusive, so as to support the interaction and shareability of information in the network environment.

10.4.7 System integration shall follow the principles of independence, modularity, consistency, extendibility, standardization and portability.

10.4.8 System integration shall be able to transfer information between different systems or modules, and to ensure data integrity and consistency.

10.4.9 System integration shall provide access interfaces and SDK for various information and support interoperation between heterogeneous databases.

10.4.10 An integrated debugging shall be carried out after the integration of system hardware and system software. Sensors, acquisition instruments, communication equipment, monitoring center equipment and data acquisition indexes of monitoring software shall meet the relevant requirements. Bugs found during integrated debugging shall be removed in time and recorded. The debugging report shall be prepared after the debugging meets the requirements.

10.4.11 The logicality, correlation and adaptability for the input and output monitoring data shall be checked against the structural characteristics of the passenger station prior to system operation.

10.4.12 Security measures (such as access control system) and a security regulations shall be implemented for the monitoring system equipment room.

11 System Acceptance

11.1 General Requirements

11.1.1 The acceptance testing of a structural health monitoring system shall consist of the acceptance testing of system hardware and system software. The system contractor shall apply for acceptance, and the company in charge of system acceptance testing shall check and fill in the acceptance record form.

11.1.2 The system hardware acceptance testing shall consist of the sensor module acceptance testing, data acquisition and transmission module acceptance testing, and hardware equipment acceptance testing of monitoring system equipment room by on-site equipment inspection and documents reviewing.

11.1.3 The system software acceptance testing shall consist of the data management and analysis module acceptance testing, alarm and assessment module acceptance testing by documents reviewing such as on-site demonstration of software function and software manual.

11.1.4 The monitoring system may be put into operation and start the trial operation, and intermediate acceptance testing of a system may be carried out at the same time after the integrated debugging for hardware and software of the monitoring system is completed and meets the provisions of the design documents and this Standard.

11.1.5 By 3 months' stable operation, the monitoring system shall meet the conditions of acceptance after qualification verification and the documents handover.

11.2 Sensor Module Acceptance Inspection

11.2.1 The acceptance inspection of sensor modules shall include on-site inspection and office documents reviewing.

11.2.2 The system contractor for structural health monitoring shall provide a list of sensors, and the inspectors shall confirm the model, specification and quantity of sensors on-site according to this list.

11.2.3 The sensors shall be provided with the manufacturer's qualification certification mark, and the model, specification, quantity and technical index shall meet the design requirements. The calibration and installation records of various sensors shall be checked.

11.2.4 The installation condition and wiring arrangement of the sensor shall be inspected, and the connector between the sensors shall be placed in the sensor protective cover or cable casing to ensure the long-term stability of the transmission line. The installation and wiring arrangement of the sensor shall not cause damage to the structure. The wire end not used for sensor connection may be placed in the sensor protective cover to keep it nice and tidy.

11.3 Acceptance Testing of Data Acquisition and Transmission Module

11.3.1 The inspection of the synchronization of different monitoring data acquisition shall meet the following provisions:

    1 The synchronization deviation of data acquisition time for the same type of monitoring variables should be less than 0.1 ms.

    2 The synchronization deviation of data acquisition time for different types of monitoring variables

should be less than 1 ms.

3 During dynamic synchronous acquisition, the synchronization deviation of each sensor shall be less than 0.1 ms.

11.3.2 The acceptance testing of the data acquisition software shall comply with the following provisions:

1 The data acquisition software shall reach real-time data acquisition and display, automatic storage, cache management, real-time feedback and automatic transmission.

2 The stable and reliable communication to the database system shall be guaranteed by the data acquisition software, the equipment configuration shall be adjusted locally or remotely, and the information shall be read through the label database or local configuration file.

3 The sensor output signals, data acquisition signals and operation status signals of transmission equipment shall be collected in real-time, the system operation status shall be monitored, and the alarm shall occur in case of abnormality.

4 The instructions for adjusting the data acquisition parameters shall be accepted and proceeded, and the process shall be recorded and backed up.

11.3.3 The following items shall be checked during the acceptance testing for data transmission software:

1 Consistency, integrity, reliability and security of data transmission, as well as system openness and extendibility.

2 Function of data compression and data unpacking recovery.

11.3.4 Each acquisition strategy shall be tested whether the sensor can correctly return the testing signal when multiple acquisition strategies are required in the design.

11.3.5 The loss rate of daily data shall not be greater than 0.5% in the automatically collected data of the system.

11.3.6 When the monitoring point is the same as or adjacent to a manual testing point, the automatically collected data by the monitoring system shall follow the same trend and pattern of the testing data collected manually at the same time, and the amount of data value changes shall be similar.

11.3.7 The monitoring data shall be stable without obvious systematic deviation.

11.4 Acceptance Testing of Data Management and Analysis Module

11.4.1 The data backup capability, as well as the consistency, integrity, reliability and shareability of the data for the data storage and management module shall be tested.

11.4.2 The data processing and analysis ability online, offline or in mixed condition shall be tested. For the online data processing, the smoothness without jamming shall be tested, for the parallel processing online and offline, mutual interference shall be tested.

11.4.3 The integrity, ease of use, security and archiving qualities of data storage and information management shall be tested.

11.4.4 The data storage precision shall not be lower than the resolution of the sensor.

11.4.5 The number and data of the sensor in the remote database shall be fully matching with the database on-site.

11.5 Acceptance Checking of Alarm and Assessment Module

11.5.1 The alarm function and the threshold for each safety grade shall be checked, and the automaticity and real-time of safety alarm shall be tested.

11.5.2 The functions of processing, display and sending of alarm information shall be verified, and information missing and false alarm are not allowed.

11.5.3 The actions against the alarms for all grades shall be checked.

11.5.4 The methods and achieved functions of structural condition assessment shall be checked.

11.6 System Acceptance Documents

11.6.1 The following documents shall be submitted for the completion acceptance of the monitoring system:

1. Application report for acceptance.
2. As-built drawings.
3. Project final accounts.
4. Records on joint review of monitoring scheme and design change request.
5. Quality certificate of materials and hardware.
6. Installation records including necessary inspection and test records.
7. Records of project quality accidents and mitigation actions.
8. System operation and maintenance manuals.
9. List of system hardware and system software.
10. Internal wiring diagram of system.
11. Integrated debugging report of system.
12. Third party test report for software.

12 System Maintenance

12.1 General Requirements

12.1.1 The system contractor shall provide the system maintenance manual, clarify the maintenance scope, train full-time maintenance personnel arranged by the monitoring system end user, and keep training record details.

12.1.2 For the end user, the management and operation ability shall be provided with.

12.1.3 The end user shall establish a maintenance regulation to carry out daily routine inspection and regular maintenance and management for the monitoring system.

12.1.4 The maintenance requirements, maintenance process, common equipment faults and repair methods for the system shall be specified in the maintenance manual.

12.2 Hardware Maintenance

12.2.1 The hardware maintenance of the monitoring system shall include the following contents:
1 Equipment running conditions of data acquisition and transmission module.
2 Equipment running conditions of data management and analysis module.
3 Equipment running conditions of alarm and assessment module.
4 Equipment running conditions of computer network transmission.
5 Equipment running conditions of monitoring center.

12.2.2 The monitoring system shall be turned on and turned off following the correct procedures and steps strictly. It is forbidden to turn on and turn off the equipment frequently, and plug in and plug out with power.

12.2.3 Daily on-site routine inspection of the system shall be carried out, the appearance of the monitoring equipment shall be inspected, and the equipment shall be cleaned, rust and ponding on the equipment shall be removed. The power cables and data cables shall be inspected to check whether the cables are damaged or aged, and whether the connection between the cables and sensors or collectors is loose. The router and the Network Interface Controller (NIC) shall be inspected to check whether they work normally, and whether the network cable connection is loose.

12.2.4 The regular inspection, maintenance and cleanness of the system equipment shall be carried out, and the equipment operation status shall be confirmed. In addition, the system error records shall be checked to eliminate the potential dangers, and to prevent and reduce the system faults.

12.2.5 The spare parts for vulnerable parts of the system shall be provided, and the damaged parts shall be repaired or replaced in time.

12.2.6 The technical related equipment or parts shall be maintained step by step under the guidance of the manuals or professional technicians from the manufacturer.

12.2.7 The troubleshooting and solutions for hardware equipment faults shall be imposed by professionals. The fault, troubleshooting process and solutions shall be recorded in detail.

12.2.8 The system maintenance shall be completely recorded, including maintenance date, maintenance director, participants, scope of maintenance, maintenance process and maintenance results.

12.3 Software Maintenance

12.3.1 The daily maintenance of software shall be carried out to ensure the monitoring data to be integral, to check whether there are error records to ensure the normal and stable operation of the system software.

12.3.2 The scope of daily checking of monitoring system software shall include the following contents:
   1 Running conditions of the data acquisition and transmission module.
   2 Running conditions of the data management and analysis module.
   3 Running conditions of the alarm and assessment module.

12.3.3 The maintenance personnel shall use the testing software to check the network transmission quality and whether packet drop occurs, and use the computer system's built-in network command to test the connection between the server and the switch.

12.3.4 The system data shall be backed up regularly to prevent data loss due to errors and defects of the system software.

12.3.5 When the software is running, the occurred errors and defects shall be managed by professionals in time.

12.3.6 Software maintenance shall be registered, and software maintenance records shall be developed to describe the background of error in detail and to record the scope of maintenance.

12.4 Emergency Maintenance

12.4.1 After typhoon, earthquake, extreme weather and other special events, special maintenance shall be carried out for the system hardware and system software, especially checking and testing the work status of various sensors installed outdoors.

12.4.2 The risk management plan shall be developed for monitoring system. The key modules and essential equipment of the system shall be maintained emphatically.

12.4.3 The emergency maintenance plan shall be developed for monitoring system. The actions shall be taken in a timely and orderly manner according to the emergency maintenance plan after a system failure warning occurs.

Words Used for Different Degrees of Strictness

In order to mark the differences in executing the requirements in this Standard, words used for different degrees of strictness are explained as follows:

(1) Words denoting a very strict or mandatory requirement:

"Must" is used for affirmation; "must not" is used for negation.

(2) Words denoting a strict requirement under normal conditions:

"Shall" is used for affirmation; "shall not" is used for negation.

(3) Words denoting a permission of slight choice or an indication of the most suitable choice when conditions permit:

"Should" is used for affirmation; "should not" is used for negation.

(4) "May" is used to express the option available, sometimes with the conditional permit.